The Wonderful World of Plants

Plants!...	2
Reaching Roots	4
Standing Stems	6
Lush Leaves	8
Now We're Cooking!	10
More About Photosynthesis.......................	12
Friends and Foes......................................	14
Glossary...	16

SCHOOL PUBLISHERS

Orlando   Austin   New York   San Diego   Toronto   London

Visit *The Learning Site!*
www.harcourtschool.com

Plants!

Plants can be many shapes and sizes. Grass plants are small, while trees may be very large. Rose bushes have flowers that smell good. Other plants don't have flowers and don't smell at all.

Plants are a little bit like you. You need certain things to stay alive. You need food, water, and shelter. You also need air and light. Plants have needs, too. Plants need air, water, soil, and light to stay alive.

Just like you, plants need air and water. Without water, a plant will lose its leaves and die. Plants take in air and use part of it to make food. Without air, plants cannot make food.

Uh-oh! Looks like someone forgot to water this plant. Without water, the plant will die.

> This plant is growing toward the light source.

Plants also need soil. Few plants can grow where there is not much soil. Very few plants can live in the desert.

Light is very important to plants. Like people, plants need light from the sun. Plants will even grow toward a light source to get enough energy. The sun's light provides energy for both plants and people.

 COMPARE AND CONTRAST Compare and contrast your needs with a plant's needs.

Fast Fact

Some plants need meat! The Venus' flytrap is covered with tickly hairs. When an insect lands on one, SNAP! Two leaves of the plant shut, trapping the insect. Yum!

Reaching Roots

Plants get nutrients from soil. **Nutrients** are the part of the soil that helps plants grow and stay healthy. So how do plants get nutrients? They use their roots. **Roots** are parts of the plant that grow underground. They take in water and nutrients from the soil. Roots also help plants stay anchored in the soil.

You can't see the roots on these flowers—they're underground. Roots get nutrients and water from the soil.

Some plants can even grow in water. The bald cypress is a tree with roots that let it grow in swampy water. Its roots form a wide circle around it. They grow deep into the soil under the water. There they can reach nutrients. The roots also take in water from the swamp.

As it gets older, the bald cypress grows "knees." These are bumps that grow out of its roots. Scientists think the knees may help the tree stand in the wet ground. They also help take in air.

 MAIN IDEA AND DETAILS Explain how the bald cypress tree's roots allow it to grow into the soil under the water.

Bald cypress trees are found in the Everglades. They can grow up to 30 meters (100 feet) tall!

Standing Stems

What happens to the nutrients and the water once they enter the roots? How do the roots move these things to the rest of the plant? That's where the stem comes in. The stem is another part of the plant that helps plants meet their needs. The **stem** grows above the ground and helps hold the plant up. The stem connects the roots and leaves. Water and nutrients move up through stems.

The tall sunflower plant has a large flower. Its thick stem makes sure enough water and nutrients can reach all the way to the top!

Fast Fact

How does the saguaro cactus survive in the desert? Its shallow roots collect as much water as possible, and its thick, waxy stem stores it between rains. The cactus can grow up to 15 meters (50 feet) tall and live 175 years!

Sunflowers need a lot of sun. The long, thick stems hold the large flowers high.

Different plants have different stems to help them meet their needs. Water thyme has stems that help it live in water. These stems can grow up to 8 meters (25 ft) long. Water thyme has small leaves. The leaves grow from the stem. A longer stem grows more leaves. Water thyme can grow stems so long that they clog lakes. Other plants and fish may die.

COMPARE AND CONTRAST Compare and contrast how stems and roots help plants meet their needs.

Water thyme came to Florida from Asia. It can take over lakes and kill native plants!

Lush Leaves

Roots are not the only way a plant gets nutrients. The part of the plant that grows out of the stem is the **leaf**. Leaves use sunlight, water, and air to make food.

Leaves are the parts of plants that take in sunlight. They are also the parts of the plant that take in air.

Banana trees are small. They grow in the shadows of larger trees. They have huge leaves that let them soak up a lot of sun, even in the shade of other trees.

A larger leaf means more room to catch some rays.

The leaves of these water lilies have a special wax coating to keep them from getting too wet.

Different plants have leaves with different shapes. Water lilies grow in water. They make food in their leaves. So, the leaves need sunlight. How do they reach it? A water lily is rooted in soil on the bottom of a pond. One long stem reaches to the top of the water. There, the stem attaches to a single, wide leaf. The leaf can get sunlight on top of the water. Because the plant has only one leaf, the leaf is very large. It can collect enough sunlight to make food. It also gets enough air for the plant.

 MAIN IDEA AND DETAILS Why does the water lily have such a large leaf?

9

Now We're Cooking!

You know that plants make food in their leaves. You also know that they use sunlight to do so. Let's find out how they turn sunlight into food!

Plants make food in a process called **photosynthesis**. For photosynthesis, plants need water, air, and sunlight. The food they make is sugar.

You have learned how plants get water. First, the roots soak up water. Then, the stem carries the water to the leaves.

Pumpkins are plants, too. The more food their leaves can make, the larger the pumpkin can grow!

This is a leaf under a microscope. You can see the holes that take in carbon dioxide.

Fast Fact

Leaves DON'T turn colors! Many plants stop making food in the fall. So, leaves stop making the chemical that turns them green. All those colors were in the leaf all year!

The air is filled with different gases. Plants need one of these gases—carbon dioxide—to make food. Every leaf has many tiny holes. The plant takes in carbon dioxide through these holes.

The plant needs one more thing to make food—sunlight. Sunlight gives the plant enough energy to make sugar. Plants have a chemical inside their leaves that turns them green. This chemical also helps leaves use the sun's energy. This energy helps the leaf make sugar from water and carbon dioxide.

 CAUSE AND EFFECT What gives plants energy to carry out photosynthesis?

More About Photosynthesis

Photosynthesis makes sugar. Plants give off oxygen into the air. Oxygen is made during photosynthesis.

Photosynthesis requires carbon dioxide, water, and sunlight to make sugar. It also makes oxygen, which we need.

Plants take carbon dioxide out of the air. They use it for photosynthesis. Photosynthesis makes oxygen that is given off through holes in the leaves.

Once the oxygen has been exhaled, it joins other gases to make up air.

Do you know what you breathe in? Oxygen! You breathe in the oxygen that plants give off. If we had no plants, we would not have the air we need to breathe.

Focus Skill **CAUSE AND EFFECT** **What effect does cutting down a forest have on air?**

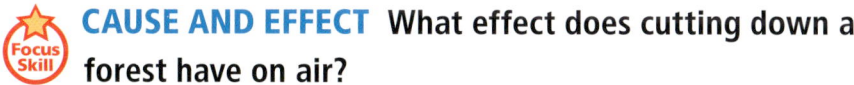

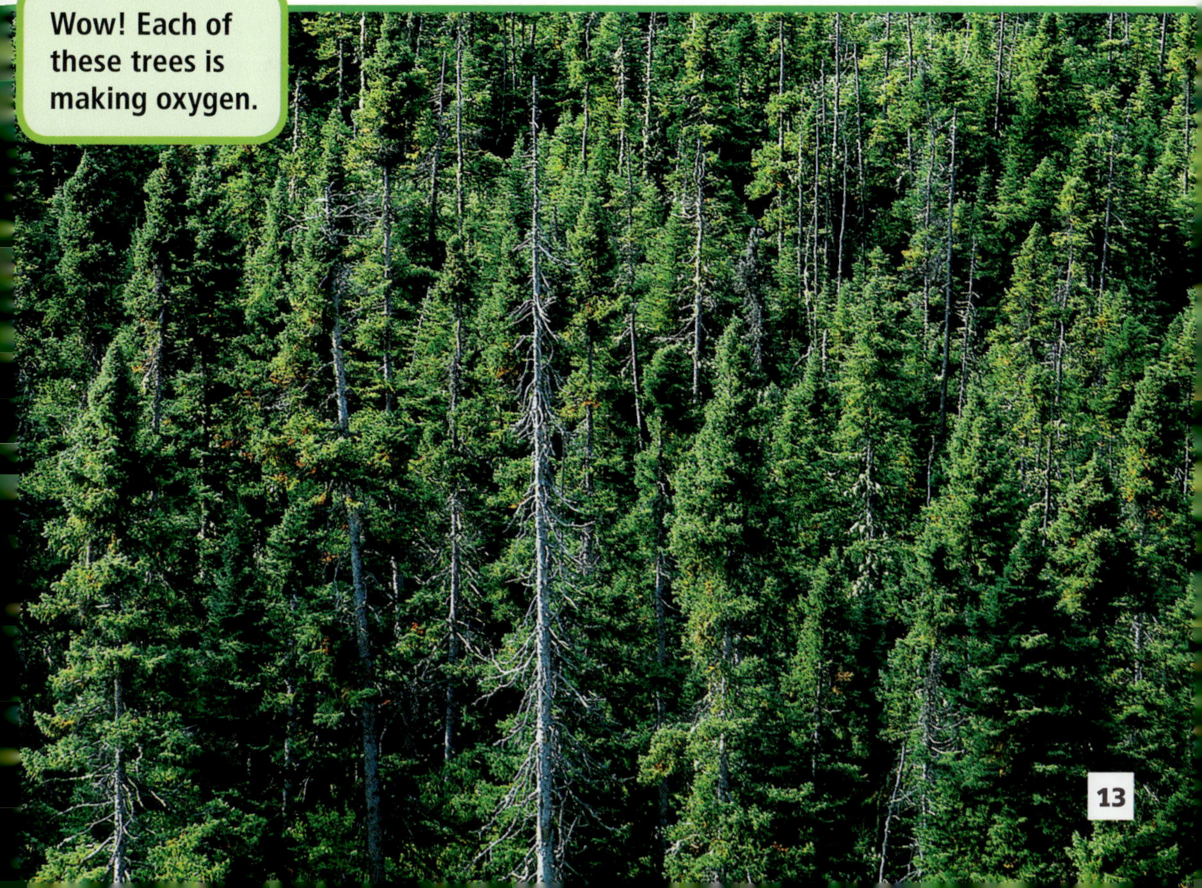

Wow! Each of these trees is making oxygen.

Friends and Foes

Plants help people in many ways. You already know that plants make oxygen. How else do they help?

People eat many plants. Fruits and vegetables all come from plants. People use the wood from trees to make houses and furniture. They also use trees to make paper.

Some plants are poisonous. If you eat them, you can die. Other plants, such as poison ivy, can give you an itchy rash if you touch them. Some plants can give people allergies.

> **Fast Fact**
>
> In ancient Greece, poisonous plants were used to put prisoners to death.

MAIN IDEA AND DETAILS Name one way plants are harmful and one way they are helpful.

This flower may look lovely, but it's deadly.

Some places have parades to celebrate flowers! Each of the floats is made using flowers.

Summary

You can find plants almost anywhere on Earth. Their roots absorb water and nutrients from the soil. The stem carries these to the leaves. During photosynthesis, plants make sugar using water, carbon dioxide, and sunlight. Photosynthesis also makes oxygen. Plants give off the oxygen, and animals breathe it in. Some plants are poisonous, but many others are used for food. Wood from trees is used to make houses. Without plants, people and animals could not survive.

Glossary

leaf (LEEF) The part of a plant that grows out of the stem and is where the plant makes food (2, 3, 6, 7, 8, 9, 10, 11, 13, 15)

nutrients (NOO•tree•uhntz) The part of the soil that helps plants grow and stay healthy (4, 5, 6, 8, 15)

photosynthesis (foht•oh•SIHN•thuh•sis) The process that plants use to make sugar (10, 11, 12, 13, 15)

root (ROOT) The part of a plant that grows underground and takes water and nutrients from the soil (4, 5, 6, 7, 8, 9, 10, 15)

stem (STEM) The part of a plant that grows above ground. It helps hold the plant up and moves water and nutrients to other parts. (6, 7, 8, 9, 10, 15)